The Earth's crust under Britain

The British Isles and the surrounding shallow seas are underlain by *continental crust* which is part of the Eurasian continental mass. Beyond to the northwest is the very different *oceanic crust* beneath the Atlantic Ocean. The continental crust under Britain is between 27 and 35 kilometres thick, as computed from the paths and speeds of earthquake waves set off by big explosions (fig 2). Since no drill hole has penetrated deeper than about 4500 metres, the composition of most of the crust can only be guessed at. It seems reasonable to assume that the deeper layers of the crust are very old and also that they should show signs of deep burial. Rocks of just this type, whose mineral constituents and crystalline textures show that they were once under high pressure and very hot, are found in the North-West Highlands of Scotland and in the islands of the Outer Hebrides. They also underlie the Rockall Bank, which is a detached part of the continent far out in the Atlantic. These crystalline rocks (fig 3) are known to be at least 2700 million years old, though probably not as old as the oldest rocks so far dated (from West Greenland, Rhodesia, Minnesota and Enderby Land in the Antarctic) which are between 3500 and 4000 million years old (the age of the Earth is

4600 million years). It can be supposed that similar old rocks form a layer running under most if not all of the British Isles. Small patches of old crystalline rock come to the surface in the southeastern tip of Ireland (fig 4) and in central Anglesey, but the next extensive outcrops after Scotland are in the Channel Islands and Brittany. The thickness of the old crystalline layer must vary a lot: it is known from the surface structure that in some places, such as the Solway Firth, younger rocks plunge deeply down while the depth to the base of the crust remains unchanged; in fact, some geologists think that the old crystalline layer may be missing altogether under southern Scotland and central Ireland. Precisely how and when the first continental crust under Britain began to form is unknown. Probably it originated towards the end of the mysterious first 1000 million years of the Earth's history by repeated remelting of volcanic lava and ash. Subsequently, portions of newer, mainly sedimentary crust have been 'welded' onto the old crystalline layer. These in turn have been reworked by further compressions and heating and injected with molten rock from deep sources. Finally, undisturbed layers of sedimentary rock have been deposited over all these older components.

3 Lewisian Gneiss, Outer Hebrides

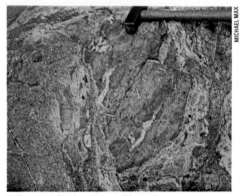

4 Rosslare Gneiss, County Wexford

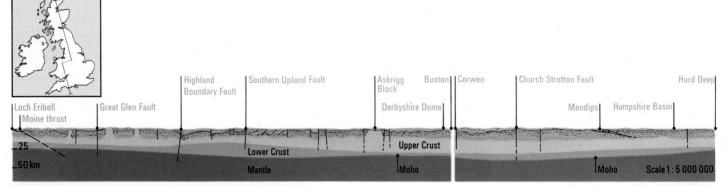

2 The Earth's crust under Britain, from data by D. Bamford and others

Plates, oceans and mountains

Most geologists accept that the outer solid Earth is made up of separate plates which are continually growing by addition of fresh volcanic rock at the mid-ocean ridges in the process called *sea-floor spreading*. Moving slowly outwards from the ridges across the ocean floor, the plates eventually sink back into the Earth's interior along *subduction zones* beneath ocean trenches, generating earthquakes and chains of volcanoes. Many mountain ranges are composed of sedimentary rocks which were deposited both in deep water in association with volcanic rocks like those found in mid-ocean ridges and volcanic island chains, and in shallow-water on continental shelves. All these rock formations are strongly folded and overthrust. This has led geologists to suppose that fold-mountain ranges originated when continental masses and volcanic island chains carried about on moving plates collided with each other after the oceanic crust separating them was totally consumed in subduction zones. These mechanisms are known as *plate tectonics*. Fig 5 a–d shows the origin of the Caledonian mountain chain in the British Isles as explained by Professor John Dewey in terms of plate tectonics. Despite the popularity of plate tectonic solutions, the classical 'ensialic' theory of mountain-building (fig 5 e, f) still finds some support. In this, thick piles of sedimentary and volcanic rock accumulate in transcontinental downwarps or 'geosynclines' supposedly located over descending currents in the Earth's interior. When the continental crust below the downwarps melts, the sediments are squeezed between the converging sides. Although this process is nowhere seen happening today, it may have happened in the more remote past. Studies of rock magnetism have shown that continental masses on opposite sides of many younger fold-mountain belts have drifted together, but similar studies of older fold-belts in Canada and Africa have failed to detect any such movement.

PLATE TECTONIC THEORY

a 600 million years

b 450 million years

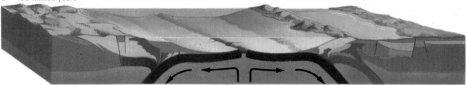

c 400 million years

d 375 million years

CLASSICAL OROGENIC THEORY

e 400 million years

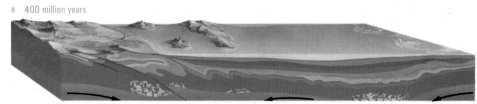

f 375 million years

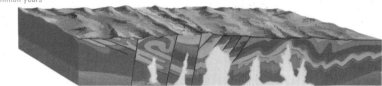

5 Origins of the Caledonian mountain belt by plate tectonics (a–d) and ensialic mechanisms (e, f)

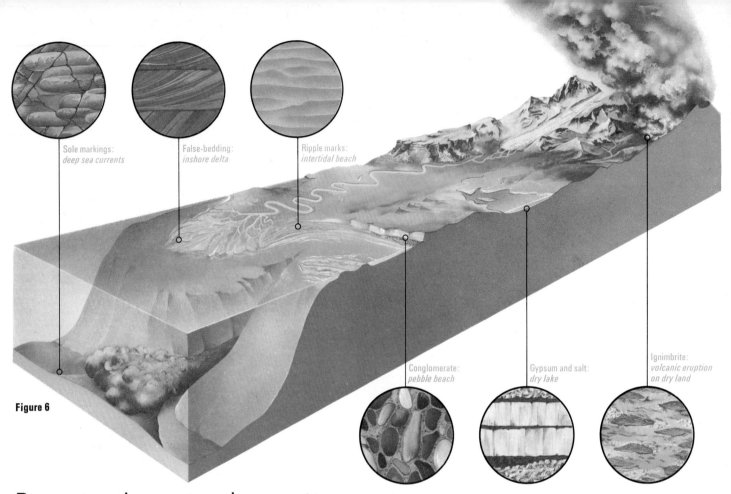

Sole markings:
deep sea currents

False-bedding:
inshore delta

Ripple marks:
intertidal beach

Conglomerate:
pebble beach

Gypsum and salt:
dry lake

Ignimbrite:
volcanic eruption on dry land

Figure 6

Reconstructing past environments

The reconstruction of ancient environments depends on a close knowledge of present-day environments such as deserts, river estuaries, deep-sea troughs, volcanic lava fields, coral reefs and ice-fields. Features diagnostic of these environments are sought in older hardened deposits of all ages. Because older rock formations are frequently deformed, weathered, eroded or otherwise altered, the full reconstruction of an environment in all its detail is a complex piece of detective work. And occasionally rocks and structures are found which have no apparent analogues in the world today.

The rounded pebbles in conglomerates (fig 6) indicate former beaches and similar 'high-energy' environments where the energy of waves or currents was capable of grinding together large stones. Ripple-marks found on bedding planes of sandstone have usually formed in shallow water while patterns of connected infilled cracks in shale or siltstone beds betray their formation in sun-dried mudflats. Peculiar lobate and fluted shapes (*sole marks*) on the undersides of greywacke-mudstone layers indicate currents presumed to be similar to the deep-water 'turbidity currents' respon-

sible for breaking submarine telephone cables. Characteristic flattened textures in certain volcanic formations show that they are *ignimbrites* formed by the welding together of red-hot particles from 'glowing cloud' eruptions: the welding can only occur on dry land. In contrast, *pillow-lavas* form only under water. Rounded sand grains in sandstones indicate deserts while the slope of dune-sandstone layers shows the direction of prevailing winds. Salt and gypsum layers result from strong evaporation in hot arid climates while glacial conditions are suggested by fossil boulder clay or *tillite*.

Building up the structure

The bedrock of the British Isles is built up from rock formations laid down in a long succession of different environments. Each environment gave place to its successor in consequence of earth movements which also affected in some degree the deposition of sediment in different parts of that particular environment. Gentle uplift caused shallowing of the sea and emergence of the sea bed as dry land, so that marine sediments are overlain by terrestrial sediments resting on a surface of erosion. Gentle subsidence flooded former land surfaces so that shallow-water marine sediments rest on terrestrial sediments. Quite small uplifts or subsidences of a hundred metres or less changed the environment radically from, say, a river floodplain to a coral sea, and *vice versa.* But the rock formations produced in these environments, however different, lie more or less conformably one layer on top of another.

At intervals in British geological history, much more powerful, mainly horizontal forces have acted on the crust, causing violent deformation of the rock layers, accompanied by heating and crystallisation of the rock on a regional scale, and followed by the injection of molten igneous rock into the folded layers. These great events, perhaps caused by collisions of continents borne along on plates, gave rise to mountainous terrains in which the folded and overthrust rock formations were subjected to rapid erosion. Within the mountains, local rift valleys and lake basins accumulated terrestrial sediments. Sooner or later, the mountains themselves were reduced to a plain which, by gentle subsidence, disappeared under the sea. The younger terrestrial and marine sedimentary layers rest discordantly on the upturned, worn-down edges of the older folded rock layers. The eroded surface on which the basal bed of the discordant layers rests is called an *unconformity* (figs 7, 8). Unconformities may be quite minor and local; the angle between the eroded beds and the overlying discordant beds may be fairly small, and the age difference between the beds may be only slight. On the other hand, unconformities generated within fold-mountain belts after strong folding followed by deep erosion are usually regional in extent and profound, slicing across major structures and reaching down to much older, deep-seated formations in the roots of the mountain chain.

Looking at the British Isles and ignoring small-scale unconformities, it is possible to recognise at least four unconformities which constitute country-wide discontinuities in the bedrock. These discontinuities separate the five major structural units from which the country is built. Fig 10 shows four of these units and, in addition, the Variscan orogenic belt, a strongly folded lateral variant of the 'older cover'. The Precambrian basement comprises two units not separately shown in Fig 10. In comparing this structural map with the geological map (fig 9), it is apparent that rocks formed in several geological periods are grouped together in each unit.

a Sedimentation

b Folding, uplift and erosion

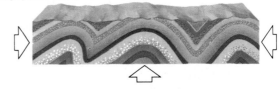

c Submergence

d Renewed sedimentation

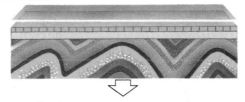

7 Origin of an unconformity

8 Unconformity, Carboniferous on Silurian, Horton-in-Ribblesdale, N Yorkshire

Figure 9

Geological map of Britain and Ireland

The map shows sedimentary rocks classified according to their age of deposition and igneous rocks according to their mode of origin. The colours are those of the international scale for geological maps. Figures indicate age in millions of years.

SEDIMENTARY ROCKS

CENOZOIC

Tertiary and marine Pleistocene Mainly clays and sands
Pleistocene glacial drift not shown — up to 65

MESOZOIC

Cretaceous Mainly chalk, clays and sands — 65-140

Jurassic Mainly limestones and clays — 140-195

Triassic Marls, sandstones and conglomerates — 195-230

PALAEOZOIC

Permian Mainly magnesian limestones, marls and sandstones — 230-280

Carboniferous Limestones, sandstones, shales and coal seams — 280-345

Devonian Sandstones, shales, conglomerates (Old Red Sandstone); slates and limestones — 345-395

Silurian Shales, mudstones, greywacke; some limestones — 395-445

Ordovician Mainly shales and mudstones; limestone in Scotland — 445-510

Cambrian Mainly shales, slate and sandstones; limestone in Scotland — 510-570

UPPER PROTEROZOIC

Late Precambrian Mainly sandstones, conglomerates and siltstones — 600-1000

METAMORPHIC ROCKS

Lower Palaeozoic and Proterozoic Mainly schists and gneisses — 500-1000

Early Precambrian (Lewisian) Mainly gneisses — 1500 to 3000

IGNEOUS ROCKS

Intrusive: Mainly granite, granodiorite, gabbro and dolerite

Volcanic: Mainly basalt, rhyolite, andesite and tuffs

0   100   200   300 kilometres

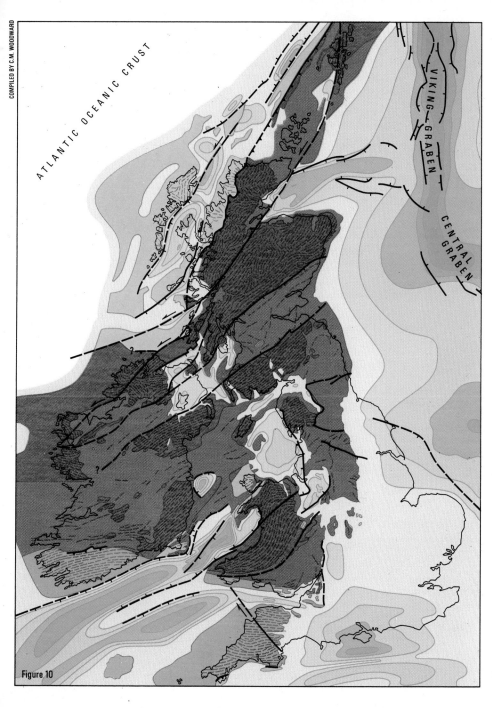

COMPILED BY C.M. WOODWARD

ATLANTIC OCEANIC CRUST

VIKING GRABEN

CENTRAL GRABEN

Figure 10

Main structural units of Britain, Ireland and surrounding seas

The map shows those groups of strata between which there are major regional unconformities, though for simplicity the unconformities within the Precambrian basement are not shown thin Lower Palaeozoic cover on the Precambrian basement is also omitted.
The Variscan orogenic belt and the 'older cover' are equivalent in age and are not separated by an unconformity.

 Younger cover: Permian to Palaeogene formations, in part Alpine-folded, plus post-Alpine Neogene and marine Quaternary

Older cover: Old Red Sandstone and Carboniferous, in part Variscan-folded, in part Caledonian-folded (Lower Old Red Sandstone)

 **Variscan orogenic belt:** Devonian and Carboniferous formations intensely folded in the late Carboniferous

 Caledonian orogenic belt: Lower Palaeozoic and late Precambrian formations folded in the early Ordovician and late Silurian

 Precambrian basement: Metamorphic and non-metamorphic rocks deformed by late Precambrian Cadomian folding in south Britain but stable since the mid-Proterozoic in north Britain

Fold-trends

 Major overthrusts

 Major faults

Basins in the younger cover

0   100   200   300 kilometres

Geological eras in the British Isles

The first mark on the British geological time-scale is 2700 million years, this being the oldest date obtained from a British rock by the method of isotope age determination. In this method, ages are computed from the ratios of radio-active isotopes of certain elements to their decay products which have steadily accumulated in the rock since it cooled or hardened. Most dates obtained in this way relate to *events* such as the crystallisation of metamorphic rocks during mountain-building or the cooling of igneous rocks. Many such dates have come from the old crystalline basement in northwest Scotland, which was stabilised 1500 million years ago. Higher up the time-scale, the ages of hardening of otherwise unaltered sedimentary rocks have been worked out, with the result that most of the unfossiliferous rocks prior to the Cambrian Period (when abundant fossils first appeared) have been placed on an internationally accepted Precambrian time-scale, shown in fig 11. Many events since the beginning of the Cambrian have also been dated by the isotope method, but the subdivision of time from 570 million years ago to the present is based mainly on other considerations. These are principally evolutionary changes in fossil faunas, in combination with successions of distinctive rock formations and the unconformities which separate them.

Post-Precambrian time, also called *Phanerozoic* time, is subdivided into the *Palaeozoic, Mesozoic* and *Cenozoic Eras.* The Palaeozoic Era is itself divided into the *Lower* and *Upper Palaeozoic,* while the Cenozoic is divided into the *Palaeogene* and *Neogene* (together making up the *Tertiary*) and the *Quaternary,* which brings us up to the present. The main subdivisions in this booklet, however, are based on the great mountain-building episodes or *orogenies,* between which more or less steady

accumulation of rock strata took place (fig 12). At the end of Precambrian time, around 600 million years ago, came the Cadomian orogeny, which affected the southern half of the British Isles. At the end of Lower Palaeozoic time – comprising the *Cambrian, Ordovician* and *Silurian Periods* – came the main Caledonian orogeny which affected almost the whole of the British Isles. The Variscan or Hercynian orogeny affected mainly *Devonian* and *Carboniferous* rocks south of the Scottish Highlands. The Alpine orogeny, affecting mainly southern England, came at the end of the Palaeogene, and the interval between the Variscan and Alpine orogenies embraced the *Permian Period* (still faunistically a part of the Upper Palaeozoic), the Mesozoic comprising the *Triassic, Jurassic* and *Cretaceous Periods,* and the Palaeogene comprising the *Paleocene, Eocene* and *Oligocene Epochs.* The post-Alpine Neogene comprises the *Miocene* and *Pliocene Epochs,* while the subdivisions of the Quaternary are based on advances and retreats of the ice sheets.

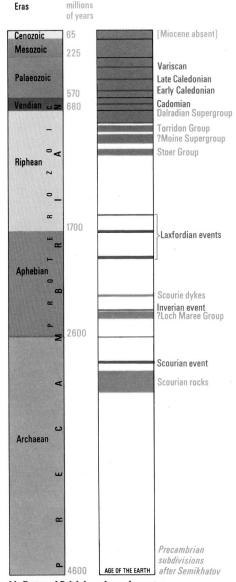

Eras	millions of years	
Cenozoic	65	[Miocene absent]
Mesozoic	225	
Palaeozoic		Variscan
		Late Caledonian
		Early Caledonian
	570	
Vendian	680	Cadomian
		Dalradian Supergroup
Riphean		Torridon Group
		?Moine Supergroup
		Stoer Group
	1700	Laxfordian events
Aphebian		Scourie dykes
		Inverian event
		?Loch Maree Group
	2600	
		Scourian event
		Scourian rocks
Archaean		
	4600	AGE OF THE EARTH

Precambrian subdivisions after Semikhatov

11 Dates of British rocks and events

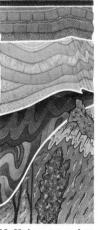

Neogene-Quaternary

Permian to Palaeogene, in part folded 25 million years ago

Devonian-Carboniferous, folded 290 million years ago

Lower Palaeozoic and older, folded 400 – 500 million years ago

Precambrian, folded at various times up to 600 million years ago

12 Main structural storeys in Britain

Precambrian time: the dark ages

The old crystalline rocks

A grey crystalline metamorphic rock of the type known as *gneiss* forms a distinctive, very barren knobbly terrain along the coast of north-west Scotland (fig 14) and in the Outer Hebrides. This rock – the so-called *Lewisian Gneiss* (named after the Isle of Lewis) – was once volcanic ash and sand deposited around a volcanic island chain possibly 3000 million years ago. The subsequent history of the Lewisian complex is shown in fig 15. Earlier igneous and metamorphic events may have affected the complex before the 'Scourian' episode of 2700 million years ago when it was buried by unknown forces to a depth of more than 20 kilometres (represented by the small 'hot' box in fig 15), where it was deformed and partially melted. After this episode, the events recorded in the fabric of the Lewisian Gneiss are all we know about Precambrian time in Britain until about 1000 million years ago, when the deposition of the oldest Torridonian beds on the deeply eroded Lewisian Gneiss signalled the beginning of widespread sedimentation lasting to the end of the Precambrian. Thus for a period of nearly 2000 million years, we know only what happened at great depths in the Earth's crust; vast thicknesses of rock, including whole mountain ranges, which formed the superstructure on the ancient basement during its repeated deformation and recrystallisation were eroded away more than a thousand million years ago in the 'dark ages' of Precambrian time.

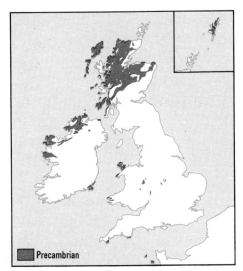

13 Precambrian rocks in outcrop

Precambrian

14 Typical Lewisian scenery, NW Highlands

?3000 million years ago
Ash, lavas and sediment deposited in a volcanic island arc

2700 million years ago (Scourian)
Very deep burial and crystallisation at high temperature and pressure

2300 million years ago (Inverian)
Uplift, squeezing and reheating with re-crystallisation

IGS D2156

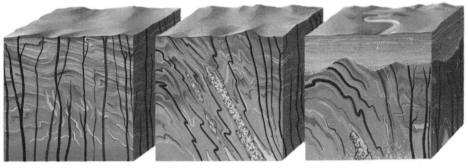

2200 million years ago (Scourie dykes)
Injection of veins of dark basaltic rock in parallel fissures

1800 million years ago (Laxfordian)
Deep burial, reheating, recrystallisation, deformation and injection of molten granite

1000 million years ago (Torridonian)
Uplift, prolonged deep erosion and final stabilisation, becoming bedrock of a river plain

15 Evolution of the Lewisian complex

The Torridon Sandstone About 1000 million years ago, the rugged Lewisian landscape began to be buried under pebbles and red sands. 200 million years later, the main mass of the Torridon Sandstone, 7 km thick, was deposited in a vast barren plain threaded by rivers rising in Greenland (then joined to Britain) and periodically flooded by a very shallow sea (fig 16). The area lay 30° south of the Equator within a supercontinent in which California was joined to Arabia! In places, the Torridon Sandstone has largely been eroded away, leaving imposing relict hills sitting on the old pre-Torridonian erosion surface (fig 17).

The Highland Trough The Scottish Highlands are built from a rock sequence 25 km thick which was intensely folded, heated and crystallised 500 million years ago. This huge sequence is divided into two parts: the bottom 10 km, called the *Moine,* forms the North-West Highlands; the top 15 km, the *Dalradian,* forms most of the Grampian Highlands and the mountains of Donegal, Mayo and Connemara (fig 18). The Moine, which rests on and is sandwiched in with the old Lewisian basement, is a monotonous sequence of banded light and dark grey rocks (fig 19) which were once reddish sand and mud deposited in large current-swept offshore deltas probably fed by the Torridonian rivers. After severe disruption by folding apparently around 730 million years ago (the evidence is conflicting), the Moine deltas gave place gradually to the very variable Dalradian marine environment of current-swept shallow waters with sandy shoals, calcareous lagoons and local deeps with muddy floors. At one stage, around 670 million years ago, a great ice-sheet (fig 20) moved over the sea-floor depositing a peculiar and distinctive boulder bed. The upper part of the Dalradian, which is of Cambrian and early Ordovician age, consists of gritty and slaty rocks deposited in a highly unstable environment of underwater volcanoes and steep submarine slopes swept by violent, avalanche-like 'turbidity currents'.

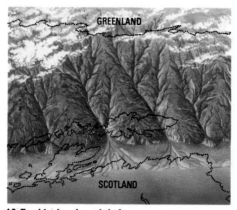

16 **Torridonian river plain from space**

17 **Relict hills of Torridon Sandstone**

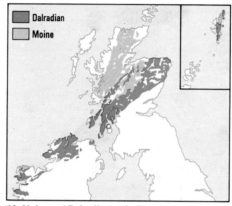

Dalradian
Moine

18 **Moine and Dalradian rocks in outcrop**

19 **Moine metasediment, current-bedded**

20 **Ice-sheet that deposited the Portaskaig-Schichallion Boulder Bed 670 million years ago**

The Mona Trough During part of the lifetime of the Highland Trough, another sea-trough (fig 21) extending across southern Ireland and North Wales was being filled with sediments and volcanic rocks. These rocks, now mainly schists (fig 22) and quartzites hardened and crystallised by burial and folding, constitute the *Mona Complex* forming much of Anglesey and parts of County Wexford (as well as eastern Newfoundland). The folding and crystallisation happened around 625 million years ago, during the deposition of the Dalradian, implying that the Mona and Highland troughs lay on opposite sides of an ocean.

The South British Volcanoes In late Precambrian time, while the Highland and Mona troughs were in being, large areas of England and Wales were dominated by volcanoes which built themselves up above the surface of the surrounding shallow waters. Many eruptions were of the explosive 'glowing cloud' type in which red-hot lava froths so fast that it blows itself to bits; the bits then hurtle down in a dense gas-cloud which deposits a blanket of red-hot ash over a wide area. There are many small outcrops of these volcanic rocks in England (fig 23) and Wales, and large areas of central England are underlain at depth by them. Fine-grained volcanic ash from the *Charnian* of Charnwood Forest in Leicestershire contains the oldest clearly visible British fossil (fig 24), a creature related to the living 'sea-pen'. In Shropshire (now Salop), a steep-sided rift valley in this volcanic terrain accumulated more than 8 kilometres of silty and sandy sediment — the *Longmyndian*. 600 million years ago, strong earth movements, the *Cadomian orogeny*, affected all of northwest Europe south of Scotland and Scandinavia. The volcanic and younger sedimentary rocks were strongly folded, in places into large structures like the Longmynd Syncline (fig 25). A rugged mountainous terrain came into existence for a few tens of millions of years before being eroded and drowned under the Palaeozoic seas.

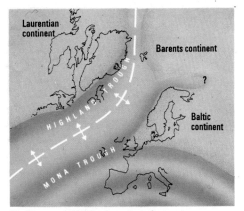

Laurentian continent
Barents continent
?
Baltic continent

21 Mona and Highland sea-troughs

22 Mona schists, South Stack, Anglesey

23 Charnian tuffs, Charnwood, Leicestershire

24 *Charnia*, a fossil sea-pen

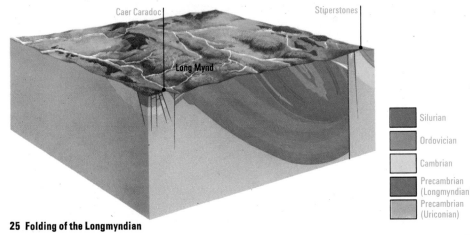

Caer Caradoc
Stiperstones
Long Mynd

25 Folding of the Longmyndian

Silurian
Ordovician
Cambrian
Precambrian (Longmyndian)
Precambrian (Uriconian)

The oceanic phase: an era of mud

Cambrian Period

Around 570 million years ago, a new era began in British geological history. The seas which flooded Britain teemed with life. For the first time there appeared, mysteriously, many animals with hard shells capable of preservation as fossils. For 150 million years, these seas deposited enormous thicknesses of mainly muddy sediment over much of Britain, making the Lower Palaeozoic Era truly an 'era of mud'. The early Cambrian seas, their wave-driven sandy margins acting like a knife-blade, cut across the tilted layers of the Torridon Sandstone (fig 28) and the old south British volcanoes. The ancient beach sands now form hard, splintery white quartzite. As the seas deepened, the sands were buried under muds in south Britain and calcareous ooze in northwest Scotland. The deep Highland Trough meanwhile accumulated muds and 'turbidite' grits which build imposing peaks like Ben Vorlich (fig 29) in the Grampian Highlands. In North Wales too, deep-water muds (now the famous Welsh slate) were deposited (fig 30), following on gritty sandstone (fig 31). Cambrian fossils from Scotland belong to a western and central American ('Pacific') fauna in contrast to English and Welsh fossils which constitute, along with those from eastern maritime states of America, an 'Atlantic' fauna. An ocean – the *Proto-Atlantic* or *Iapetus Ocean* – may have separated the two faunas.

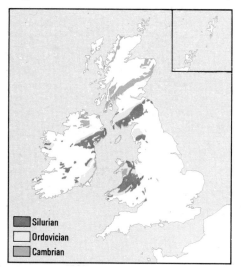

Silurian
Ordovician
Cambrian

26 Lower Palaeozoic rocks in outcrop

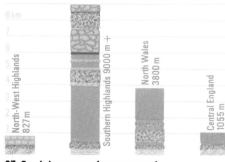

27 Cambrian successions compared

28 Cambrian discordant on Torridonian, Ben Eighe

30 Cambrian slate, Dinorwic, Gwynedd

29 Ben Vorlich, Cambrian metasediments

31 Rhinog Grit scenery, Harlech, Gwynedd

Ordovician Period

At first, little changed: carbonate ooze was still being deposited in the far northwestern 'Pacific' province while muds continued to accumulate in the Highland Trough. From Anglesey to southeast Ireland, an 'Irish Sea Landmass' rose out of the water and Cambrian sediments were eroded from it, but it was soon under the sea again. Then suddenly, around 500 million years ago, sedimentation stopped in the Highland Trough and the northwest; the Earth's crust from Shetland through Scotland, Ireland and Newfoundland was subjected to powerful stresses which folded and collapsed the Moine-Dalradian sedimentary pile; at the same time, the region was heated from below and the rocks crystallised. In south Britain at this time, the disturbances to sedimentation were minimal and limited to local block uplifts, though there was considerable volcanic activity. The separate halves of the British Isles lay in tropical latitudes between 20°S and 30°S. The South Pole was in what is today the western Sahara. The seas in north Britain were probably warm and flanked by deserts while those in south Britain were cooled by cold ocean currents flowing up from the Ordovician Antarctic (fig 32).

The Ordovician sea floor in England, Wales and Ireland was sinking at very variable rates; in consequence the amount and type of sediment deposited varied enormously from place to place. The distant, offshore sediments were mainly muds (which hardened to shales and mudstones) and muddy sands deposited from submarine avalanches (which hardened to bedded grey sandstones called *greywackes*). In shallow-water, inshore areas in south Shropshire (Salop) and Ayrshire (fig 33), sandy, pebbly, muddy and calcareous sediments were deposited in patches on

conjectural currents
——→ cold
——→ warm
——→ movement of continents

32 Ordovician continents and oceans

33 Mid-Ordovician geography 450 million years ago

the sea floor. These sediments (fig 34) contain abundant fossils, especially shells called *brachiopods* and crawling arthropods called *trilobites*. The offshore muds yield puzzling fossils called *graptolites*, which were a kind of free-floating colonial animal. To give some idea of the huge variations in thickness of sediment deposited during the same time interval, 6 metres of black graptolitic shale in the central Southern Uplands (fig 35) are equivalent in age to 3000 metres of greywacke in south Ayrshire. Early Ordovician brachiopods and trilobites cluster into distinctive 'American Realm' (Scotland and northwestern Ireland) and 'European Realm' (southeast Ireland, England and Wales) faunas. As in the Cambrian, an ocean may have separated them, even after the folding and uplift of the Highland trough as the 'Eocaledonia' landmass (fig 33). Later on, the ocean began to contract and an increasingly free exchange of faunas took place. By the end of the Ordovician, a unified cosmopolitan fauna had been established.

Ordovician volcanic activity was extensive and prolonged, breaking out at the beginning of the Period in the Southern Uplands, Wales (Cader Idris and St Davids district) and South Mayo (fig 36), and after a pause, building up to a crescendo halfway through, when the thick volcanic piles of Snowdonia (fig 37), the Lake District, Leinster and Tyrone were formed. Some eruptions, and more especially the earlier ones, were submerged 'shield' volcanoes shaped like huge upturned saucers, erupting relatively fluid *andesite* lava. Others, in Snowdonia, the Lake District and Leinster, were partly above water and erupted viscous, gas-rich rhyolite lava which often exploded in glowing clouds. These blew out over the exposed mudflats, depositing 'ash-flows' whose particles fused together to form massive *ignimbrites*. Huge quantities of ash fell in the sea, to be re-sorted by currents; the volcanoes themselves were eroded by the waves (fig 38) and buried under sands and muds.

34 Benan Conglomerate, Girvan, Ayrshire

35 Graptolitic shales, Moffat, Dumfries-shire

36 Volcanic hills, Lough Nafooey, Galway/Mayo

37 Volcanic rocks, Dinas Cromlech, Llanberis Pass

38 Reconstruction of eroded Ordovician volcano, near Builth Wells, Powys

Silurian Period

Although the Proto-Atlantic ocean had narrowed considerably, deep water persisted over many parts of the British Isles for much of the Silurian Period. In the early Silurian, the marginal shallow seas in the Welsh Borderland and central England were obliterated by fracturing and block uplifts, but elsewhere deposition proceeded without interruption. In the Southern Uplands and across into central Ireland, black graptolitic muds (now hardened to shale) were still being deposited extremely slowly on a longitudinal submarine ridge, as in the Ordovician. Similar 'condensed' mud sequences were deposited in the Lake District and North Wales, while in west-central Wales, a trough in the sea floor received copious quantities of greywacke sand (fig 39). About a third of the way through the Silurian Period, around 425 million years ago, the sea flooded over the land area in the Welsh Borderland. Simultaneously in the Southern Uplands, coarse greywacke sands derived from a newly-elevated landmass, 'Cockburnland' (fig 40), began to spread southwards across the former submarine ridge in the Southern Uplands. By the late Silurian, a great mass of greywacke sands and muds up to 8 kilometres thick was deposited in the south Southern Uplands and nearly 5 kilometres of similar sediment covered the Lake District. The sediments in the Southern Uplands may have been telescoped together during subduction of the Proto-Atlantic oceanic crust. In the shallow shelf-sea in the Welsh Borderland, the water was at times clear enough to allow the growth of coral reefs (fig 43) which supported a profuse and varied fauna (fig 41). At the same time and especially in northeast Wales, avalanche greywackes and muds with fantastic convolutions caused by slumping (fig 42) indicate steep, unstable submarine slopes and deep water. Towards the end of the Silurian Period, the by now much reduced sea basin gradually silted up altogether and the youngest Silurian deposits were formed on a land surface.

39 Aberystwyth Grits: graded greywackes

40 Mid-Silurian palaeogeography

lowland

E O C A L E D O N I A ?

Cockburnland

graptolite muds

turbidites

?

turbidites

shallow water with reefs

graptolite muds ?

?

?

41 Fossiliferous Wenlock limestone

42 Slumped beds over normal strata, Llansannan

43 Reconstruction of Wenlock patch-reefs with sea-lilies and corals

The Caledonian mountain-building

The subsidence of the deep Highland trough more than 800 million years ago was the beginning of a lengthy process which culminated 400 to 500 million years ago in the two main mountain-building episodes referred to as 'Caledonian'. The whole process of subsidence and deposition, deformation and uplift, with associated igneous activity, is called the *orogenic cycle.* The Caledonian orogenic cycle had an early and a late stage. The early stage was the accumulation of 25 kilometres of sedimentary and volcanic rocks in the Highland trough which were intensely deformed and crystallised in the early Ordovician around 500 million years ago. The late stage began with the invasion of the Cambrian seas in south Britain 570 million years ago, leading ultimately to the main folding of the vast accumulations of Lower Palaeozoic muddy sediment and volcanic rock near the end of the Silurian about 415 million years ago. The folded rocks were uplifted and became mountainous terrain capped by volcanoes around 400 million years ago. Mild folding of the thick beds of detritus eroded from the mountains – the Old Red Sandstone – took place, followed by the accumulation of more detritus in lake basins, but that was virtually the end of the cycle. The folded tract, or 'Caledonides', runs NE–SW across the British Isles (fig 44); its northwestern edge is a huge overthrust but the southeastern margin is gradational and coincides with vertical fractures.

The Early Caledonides comprise the folded and crystallised (or *metamorphic*) Moine and Dalradian successions (fig 45) together with recrystallised Lewisian Gneiss, making up the Caledonian Metamorphic Zone (fig 44). In Scotland, this zone terminates on the northwest in the Moine Thrust Belt in which Lewisian, Torridonian and Cambrian rocks are strongly dislocated ahead of a huge overthrust mass of crystalline Moine rocks. The southern limit is a deep fracture, the Highland Boundary Fault. In Ireland, metamorphic Dalradian whose folding is pre-Ordovician is exposed in Connemara south of the Highland Boundary Fault. The internal structure of the metamorphic zone is exceedingly complicated; the rock formations have been folded repeatedly, never less than twice and up to eight times, though not all in the same place. Rock successions are juxtaposed in the wrong order or cut out along *slides* which are themselves folded. The earliest structures are gigantic recumbent folds, the largest of which – the so-called *Tay Nappe* (fig 46) – extends from Aberdeen to Donegal. Upside-down Dalradian strata in its inverted underside are exposed along the south side of the Grampian Highlands, and the crest with 'parasitic' minor folds (fig 48) faces down into the ground as a result of later folding. The metamorphic crystallisation is strongest around large tracts of partially melted, 'granitised' rock known as *migmatite.*

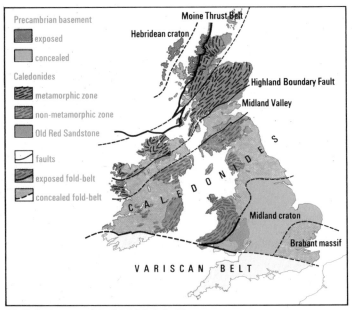

44 Main elements of the British Caledonides

45 Folded Dalradian metasediments, Kerrera, Oban, Argyllshire

The Late Caledonides are the folded slate, shale, mudstone, greywacke and volcanic sequences of Lower Palaeozoic age in the Caledonian Non-metamorphic Zone (fig 44). Each of the mainly mountainous districts has its own characteristic structure, but some features are common to all. Up to four successive foldings can generally be seen, though how far they are separate in time is controversial. Each folding has produced, particularly in muddy rocks, a laminated structure called *slaty cleavage* which in fine-grained rocks makes good-quality roofing slates. A younger cleavage may intersect an older cleavage and both may intersect the original sedimentary bedding. In the Southern Uplands, large-scale movement on inclined fractures (fig 47) has repeatedly pushed up older rocks, so cancelling out the effect of the northward inclination of the strata themselves, which is to bring on younger rocks to the north. In the Lake District and North Wales, much argument centres on whether earlier large-scale folds and related slaty cleavages in Ordovician rocks are actually of Ordovician age, perhaps related to volcanic upheavals, rather than late Silurian (fig 49). In South Mayo (fig 50), the Silurian rests unconformably on folded early Ordovician rocks.

In all the Caledonian-folded districts except Wales, extensive melting of the deep continental crust resulted in the rise of large masses of molten granite. Many granite masses did not reach the surface until the ensuing Devonian Period, when they forced their way into lava piles poured out over the eroded Caledonian mountains. In Scotland, some granites have been emplaced by filling the space above and around a subsided cylinder of metamorphic rock and overlying lava (fig 51). The thickest beds of detritus, in the Scottish Midland Valley, were buckled and eroded in the mid-Devonian. Large fractures appeared in the Highlands and Donegal, along the largest of which – the Great Glen Fault – the Northern Highlands has moved at least 105 kilometres southwestwards relative to the Grampian Highlands.

46 The Tay Nappe

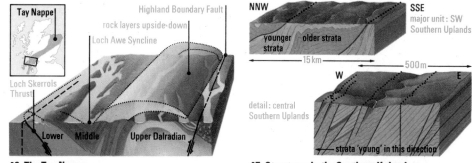

47 Structures in the Southern Uplands

48 Downward-facing folds, Birnam, Perthshire

49 Folded Silurian grits, Shap, Lake District

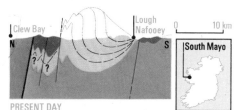

50 Evolution of the South Mayo trough

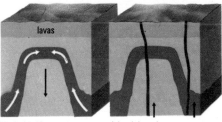

51 Mechanism of 'cauldron subsidence'

Old Red Continent, coral seas and coal forests

Devonian Period

The closing of the Proto-Atlantic ocean placed Britain in an equatorial position as part of 'Euramerica', one of the three large continents on the Devonian globe (fig 53). Most of Britain formed part of a new land area, the 'Old Red Continent', and consisted of a northern mountainous region passing southwards into a vast alluvial plain bordered by the open *Devonian Sea,* whose northern shoreline lay roughly along a line joining London and Bristol (fig 54), and which extended from Cornwall to the Rhineland and beyond. In north Devon, shoreline deposits of shallow-water marine muds and limestones alternated with non-marine delta sands derived from rivers flowing off the Old Red Continent. In south Devon and east Cornwall, thick muds, now shales and slates (fig 56), were deposited in subsiding basins while fossiliferous limestones, including coral reefs, formed on the shallower intervening ridges. In the deeper waters of west Cornwall, dark muds accumulated along with coarse sands called 'turbidites' which avalanched off the steep slopes of a landmass rising in the English Channel area. Underwater volcanoes erupted large volumes of 'greenstone' lava in many areas, particularly in the deeper waters where the action of cold water on the hot lavas produced pillow-shaped structures. This volcanic activity became more intense as the Devonian progressed.

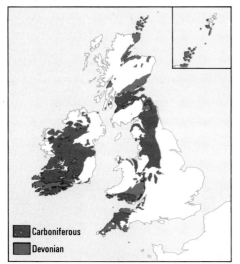

52 Devonian and Carboniferous rocks in outcrop

Carboniferous
Devonian

53 Devonian continents and oceans

54 Britain from space 370 million years ago: Old Red Continent and Devonian Sea

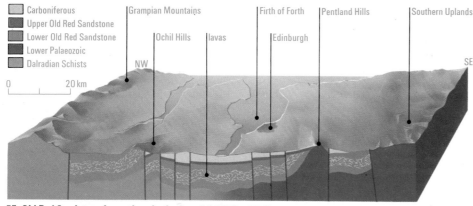

Carboniferous
Upper Old Red Sandstone
Lower Old Red Sandstone
Lower Palaeozoic
Dalradian Schists

Grampian Mountains
Ochil Hills
lavas
Firth of Forth
Edinburgh
Pentland Hills
Southern Uplands

NW
SE

0   20 km

55 Old Red Sandstone formations in the Scottish Midland Valley

56 Devonian Morte Slates, North Devon

57 Old Red Sandstone, Brecon Beacons, Powys

58 Macgillycuddy's Reeks, County Kerry

59 Caithness Flagstones

The Old Red Continent is named from its characteristic sedimentary deposit, the *Old Red Sandstone,* so-called in distinction to the *New* Red Sandstone which lies *above* the Coal Measures. The climate was hot and wet; primitive fish abounded in shallow lakes and plants colonised the land. By the end of the Devonian, real forests had formed and amphibians had evolved from air-breathing fishes. North of the Devonian Sea, a number of subsiding basins lay within a vast coastal plain extending from London to southern Ireland and threaded by a river system like the present Colorado Delta. These meandering rivers deposited repeated sequences of pebble-beds, sands and silts (fig 57) which, after 10 to 15 million years of Devonian time, were uplifted and eroded in the last phase of the Caledonian orogeny. Deposition resumed and included a brief northward advance of the sea. Great thicknesses of Old Red Sandstone accumulated, particularly in southern Ireland, where 7 kilometres of fine red silt was deposited (fig 58). Between the Scottish Highlands and the Southern Uplands, and extending into Northern Ireland, lay a large elongated basin into which torrential streams carried vast quantities of pebbles and sands. Temporary lakes appeared and chains of volcanoes poured out lavas and ash. These rocks, the Lower Old Red Sandstone, were then folded and let down between the Highland Boundary Fault and the Southern Upland Fault, creating the present-day Midland Valley (fig 55). The younger Upper Old Red Sandstone was deposited on the eroded earlier rocks by large rivers feeding a central lake.

In north-east Scotland and the Orkneys, sands and muds poured into hollows while pebble beds and screes spread out from the mountains in great fans. After uplift and erosion the area was submerged under a great landlocked body of water teeming with fish – *Lake Orcadie* – where repeated sequences of limy muds and sands were deposited (fig 59). The water was very shallow, locally and intermittently drying up and killing the fish.

19

Carboniferous Period

At the beginning of the Carboniferous, when southern Britain lay across the Equator, a warm shallow sea like parts of the Caribbean today advanced quietly northwards, covering the Old Red Continent as far as the Scottish Highlands. Some large areas remained as islands while the flooded areas steadily subsided (fig 60). The Carboniferous Period is divided into three Epochs named after European districts: the Dinantian, Namurian and Westphalian.

Dinantian Epoch

Though the most characteristic product of this Epoch is the *Carboniferous Limestone* (fig 64), many other kinds of rock also formed in various different environments. In the deeper sea-trough across southwest England, little sediment accumulated, but submarine volcanoes erupted basalt lavas and siliceous exhalations from which was deposited a flinty rock called *chert.* In southern Ireland, black muds and grits accumulated south of a line from Kenmare to Cork while Carboniferous Limestone was deposited in shallower water to the north (fig 61). Over the rest of the British Isles, several large basins developed in which muds mainly accumulated while true Carboniferous Limestone was deposited in shallower water both between the basins and bordering the islands. Typically the Limestone contains stem fragments of sea-lilies or *crinoids* (figs 62, 63), brachiopod shells and corals, but dolomitic, oolitic and reef varieties formed with fluctuations in sea-level. In Cumbria, delta muds and sands derived from the northern hills filled a large trough; later on, the deltas were repeatedly flooded, giving rise to cyclic deposition of limestone, mud, sand and coal. In the Scottish Midland Valley, great volumes of basalt lava poured out around Glasgow (fig 65) while muds rich in algae and forest plant remains (now the famous oil shales) accumulated in quiet lagoon-like waters around Edinburgh.

20

60 Britain from space 330 million years ago: islands in the early Carboniferous sea

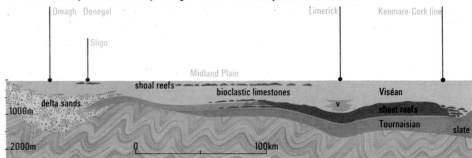

61 The Irish Dinantian, with lateral variations, resting on folded Caledonian basement

62 Sea-lily 'garden'

63 Crinoidal Carboniferous Limestone

IGS T246

Namurian Epoch

Earth movements ultimately restricted the warm Dinantian seas and filled the basins with land-derived mud and sand. Deeper water persisted in southwest England, in which great quantities of 'turbidite' sediment were deposited (fig 66). On the northern shore, in South Wales and central-western Ireland, lay a coastal plain with extensive mudflats and river deltas. In the central Pennines, marine muds accumulated in a large basin repeatedly invaded by massive river-deltas formed of sands which now constitute the familiar *Millstone Grit* (fig 67). When the basin became choked with sediment, cyclic deposition ensued as in the Dinantian Epoch but with marine muds instead of limestone. Cyclic deposits also accumulated in the shallower waters over northern Britain with outbursts of volcanism in Scotland, but towards the end of the Namurian Epoch, river deltas established themselves throughout.

Westphalian Epoch : the Coal Measures

The progressive erosion of the Caledonian mountains and infilling of the basins between them meant that by early Coal Measures time much of Europe and eastern North America were covered by vast swamps and mudflats almost at sea level and threaded by wandering streams. Dense tropical forests grew in which spiders, dragonflies, lizards and amphibians flourished. The forests were continually renewed despite repeated drowning under the sea, and on dying the rotting vegetation changed first to peat and then, after burial under great thicknesses of mud and sand, to coal. As in earlier times, cyclic deposits accumulated but were now much more widespread. A coal seam was the end-result of each cycle (fig 68) and the deposits were built up by repetition of the following events :

1 Invasion of the swamp by the sea, with deposition of marine muds.

2 Silting up of the shallow sea, conversion to a river delta and deposition of non-marine muds and sands.

3 Creation of swamp, growth of vegetation and accumulation of peat.

4 Invasion by the sea; drowning of the coal forest.

This environment persisted fairly uniformly over most of Britain apart from the shallow basin over southwest England where thick delta sands accumulated, but towards the end of Coal Measures time, the sea withdrew almost totally. North of the Wales-Brabant Island, coal swamps were severely restricted and the rocks became redder as the climate became drier. To the south, in Bristol and South Wales, coal swamps persisted but were invaded by thick sands – the 'Pennant Sandstone'– laid down by a large river flowing off emerging land areas to the south.

64 Carboniferous Limestone scarp, Llangollen

66 Namurian 'Crackington Formation'

65 Basalt lava scarp, Gargunnock Hills, Stirling

67 Millstone Grit, Stanage Edge, Derbyshire

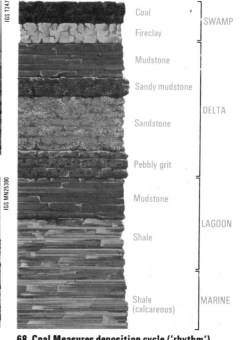

Coal — SWAMP
Fireclay
Mudstone
Sandy mudstone — DELTA
Sandstone
Pebbly grit
Mudstone — LAGOON
Shale
Shale (calcareous) — MARINE

68 Coal Measures deposition cycle ('rhythm')

The Variscan mountain-building

About 280 million years ago, the region of the British Isles was subjected to strong compressive forces which buckled the Devonian and Carboniferous strata and the older rocks on which they rest, replacing the luxuriance of the Coal Forests with the hot, arid 'New Red Desert' and generating large quantities of molten igneous rock. This was the *Variscan* or *Hercynian Orogeny,* named after localities in Germany. The main orogenic belt, marked by intense deformation, runs through southwestern England and Southern Ireland (fig 69). Its northern margin – the so-called Variscan Front – is a belt of folded and overthrust Old Red Sandstone and Carboniferous Limestone running from the Mendips to southwest Wales (figs 70b, 73). North of the Variscan 'Front' the same rocks are more gently folded in directions determined by structures in the underlying basement. Many difficulties are met in attempting to explain

Variscan movements in terms of plate tectonics. Although sea lay over southern Britain during the Upper Palaeozoic, it was a relatively shallow sea over continental crust with no evidence of either oceanic crust or a subduction zone (p 3). A northward-dipping subduction zone could have lain much further south in southern Europe; possibly a 'Japan Sea' came into existence as south Cornwall moved away from the rest of southwest England during part of Upper Palaeozoic time, just as Japan has moved away from the Chinese mainland, to be re-attached in the final plate collision at the climax of the Variscan orogeny. Alternatively, plate tectonics may have been subordinate and Variscan effects were perhaps largely produced by the upward movement of vast quantities of granite magma, as happened in southwest England towards the end of Carboniferous time (fig 72).

a. Manchester   Lancashire Coalfield   Pennine Anticline   Holme Moss

b. V a r i s c i d e s   Caledonides
Tenby

70 Variscan folding in (a) the Pennines (b) Pembrokeshire

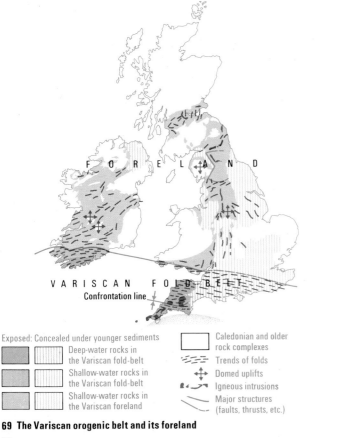

F O R E L A N D

V A R I S C A N   F O L D - B E L T
Confrontation line

Exposed: Concealed under younger sediments
Deep-water rocks in the Variscan fold-belt
Shallow-water rocks in the Variscan fold-belt
Shallow-water rocks in the Variscan foreland
Caledonian and older rock complexes
Trends of folds
Domed uplifts
Igneous intrusions
Major structures (faults, thrusts, etc.)

69 The Variscan orogenic belt and its foreland

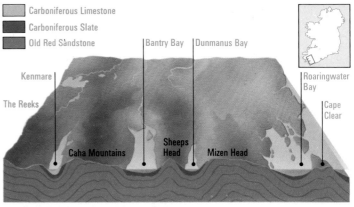

Carboniferous Limestone
Carboniferous Slate
Old Red Sandstone

Kenmare
The Reeks
Bantry Bay   Dunmanus Bay
Roaringwater Bay
Cape Clear
Caha Mountains   Sheeps Head   Mizen Head

71 Variscan folding, Southwest Ireland

Southwest England Here the Variscan disturbances were most severe. A large structural depression was formed in which the rocks were intricately folded and faulted. At depth, intense folding and refolding (fig 75) was accompanied by mild metamorphic recrystallisation which has given the Devonian rocks a slaty cleavage. Nearer the surface, the uppermost Carboniferous rocks were folded into flat-lying to upright 'concertina' folds. Two successive north–south compressions occurred, the initial *north*-facing folds meeting the later *south*-facing folds in a 'confrontation' zone (fig 74). As the compression waned, the Cornubian granite batholith rose into the rocks, producing a rich copper-tin mineralisation (fig 72).

Northern Britain North of the Variscan Front the rocks are more gently folded, the direction of the folds being determined by structures already existing in the older rocks below, for example the north–south Pennine uplift (fig 70a) over the northward prolongation of a large basement fracture, the Malvern Axis. Coal-bearing rocks were preserved in huge downfolds which erosion has separated into the present-day coalfields. An intricate system of fractures was mineralised in the Peak District and northern Pennines, and the Great Whin Sill of igneous dolerite was emplaced, traversing northern England for almost 300 kilometres. Dykes and sills of similar rock were injected in the Midland Valley of Scotland.

Ireland The intensity of Variscan folding in Ireland increases progressively southwards. The northern folds are mainly broad uplifts and basins with a NE–SW trend inherited from the Caledonian basement. On approaching a belt of dislocation running eastwards from Killarney, a 'frill' of smaller folds, mainly affecting shaly rocks, is superimposed on these gentler uplifts and basins. South of the belt, which constitutes the Variscan Front, is a series of closely spaced parallel folds running WSW which have a pronounced effect on the scenery. The long fjord-like inlets of southwest Ireland (fig 71) are eroded out of downfolds of soft Carboniferous slate with upfolds of harder Old Red Sandstone forming the intervening peninsulas.

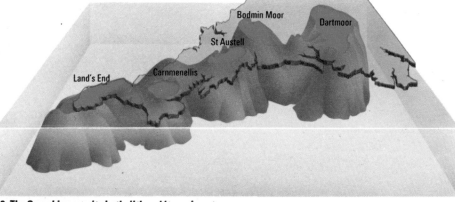

72 The Cornubian granite batholith and its main outcrops

73 Overthrust fold, Broadhaven, Dyfed

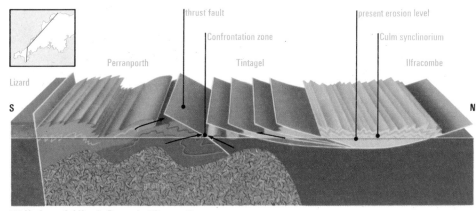

74 Variscan folding in Devon and Cornwall

75 Refolded folds, Boscastle, Cornwall

Baking deserts and tropical seas

Permian and Triassic Periods

During the 250 million years following the upheaval of the Variscan mountains, the Atlantic Ocean was formed and the dinosaurs lived and died. At first, 280 million years ago, Britain lay just north of the Equator, part of the 'New Red Desert' within the vast Pangaea Supercontinent which later split up into the present-day continents. In the first thirty million years of the Permian Period, the desert lands became increasingly arid as erosion cut into the hills. Dune sand-seas covered the plains (fig 79).

Then, suddenly, low-lying desert areas were invaded by ocean water to form warm, inland seas (fig 78) : the largest, the tideless *Zechstein Sea,* stretched from Yorkshire to Poland. For twenty million years these seas repeatedly dwindled and reflooded, leaving cyclic layers of salt (fig 77), shale and limestone (fig 80), finally giving way to monotonous desert. At the start of the Triassic Period, boulders and gravel were strewn along southern rift valleys by powerful northward-flowing rivers which spread sand

and silt – the *Bunter Sandstone* – across the Midlands. The crust on either side of the Pennine–London uplands had been subsiding gradually, but the landscape became increasingly featureless as the rifting movements died down and the hills were eroded flat. Red dust and salt pans – the *Keuper Marl* (fig 81) and *Keuper Salt* – were the empty world of the ancestral dinosaurs. Then, 200 million years ago, life-giving tidal seas spread over Britain from the south.

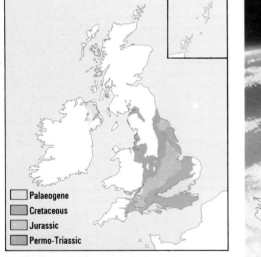

Palaeogene
Cretaceous
Jurassic
Permo-Triassic

76 Permian to Palaeogene rocks in outcrop

77 Zechstein Sea salt (quarter size)

IGS T248

78 Britain from space 250 million years ago : desert lands and the Zechstein Sea

24

79 Permian dune-sandstone, Dawlish, Devon

Jurassic Period

For over 150 million years following the Triassic Period, shallow seas repeatedly overran the land as the underlying continental crust slowly but intermittently warped and rifted during the gradual opening out of the Atlantic. An ever-changing pattern of land, sea, swamps and lakes is recorded in the varied rock formations so characteristic now of the rolling hills and wide plains of eastern and southern England. By the start of the Jurassic, the area of Britain had drifted away from the hot, dry latitudes to about 40° north of the Equator where the climate was possibly one of humid, monsoon-drenched summers. Smoothed-down remnants of the old desert uplands were now islands covered with ferns and trees, set in warm, life-giving shallow seas, and inhabited by small mammals scurrying in the shadow of dinosaurs and pterosaurs. The seas were dominated by plesiosaurs, ichthyosaurs and sea-crocodiles in company with ammonites, belemnites and other invertebrates (fig 82). Uneven subsidence of the crust ensured a variety of sea-bed conditions. In deeper water, dark muds were deposited which became the *Lias* clay (fig 83), while more limy and iron-rich sediments were deposited in shallower waters with sandy shores.

80 Magnesian Limestone, near Doncaster, Yorkshire

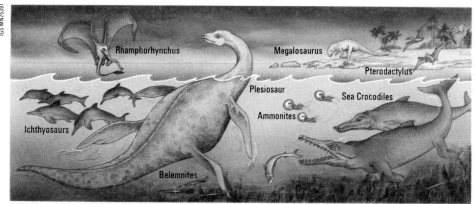

82 Life around Jurassic shores 165 million years ago

81 Keuper Marl, Aust Cliff, Avon

83 Lias cliffs, Watchet, Somerset

84 Cotswold limestone, Doulting, Somerset

Through mid-Jurassic time, the North Sea area continued to subside as renewed tensions caused the continental crust to rift. Sediments accumulated rapidly over the area, filling up the rift valleys while volcanoes towered above delta lagoons, estuaries and even coal-forming forests (fig 85). To the southwest was a shallow, warm sea in which were laid down the limestones that now form the Cotswolds (fig 84). The flat plains of the south and east Midlands owe their existence to the return of muddy seas 160 million years ago. After ten million years these seaways were narrowing. Dense 'blooms' of plankton gathered seasonally to die and sink in the stagnant mud; subsequent deep burial and natural distillation of the

resulting oil shale in the North Sea area has driven off petroleum, which was trapped in overlying porous sand formations. Warping of the crust accompanying early stages in the opening of the North Atlantic was slowly tilting Britain to the southeast while at the same time world sea-levels were falling. Land areas increased in the northwest and the seas divided (fig 86). The mud and sands of the *Volgian Sea* stretched from Lincolnshire to Moscow. South of the 'London Island' was the quite different, possibly warmer, *'Portland Stone Sea'* with its limy deposits. That the two sea areas coexisted was established from fossil ammonites, which can give a precise indication of the relative ages of widely separated rock outcrops.

Cretaceous Period

During the Cretaceous the new North Atlantic Ocean pushed America and Greenland apart (fig 93b), bringing to a halt the tensional rifting around Britain. The area meanwhile drifted back towards the Equator from a latitude close to that of the present day. By the end of the Jurassic, the southern seas (fig 86) had shallowed to salt-lakes, lagoons and forested mud-swamps trodden by dinosaurs. The climate gradually altered from semi-arid to warm and seasonally wet. As the surrounding lands rose in a series of sharp uplifts, torrential floods periodically swept across a mud-plain clad in 'horsetail' bush grazed by herds of Iguanodon dinosaurs. Eventually, after more than twenty million years, this *Wealden* plain was invaded by the *Lower Greensand Sea.* In the *Volgian Sea* to the north muds and sands had been deposited since late Jurassic time. 100 million years ago, the 'London Island' was at last buried under thick mud when it was submerged by the *Gault Clay Sea* (fig 87). Near-shore sands spread into this sea from the granite hills in the southwest. As the crust moved again, the new rocks were warped and fractured. The new ocean to the west started to widen rapidly (fig 93); world sea levels continued to rise inexorably until there was little if any land left where Britain now lies. Seventy million years ago, the age of seas had reached its peak; Britain was blanketed by chalk mud largely made up of the fragmented hard parts – *coccoliths* – of planktonic algae (fig 88). Up to 500 metres of the resulting peculiar limestone – the Chalk – make up the dry downland and white cliffs of south and east England (fig 91). The Chalk and, to the north, muds and sands, buried the old, still sagging, rifts (fig 98) which were no longer under the tensions of incipient continental splitting. Many forms of life were disappearing; the dinosaurs diminished to extinction amid a modern-looking scene with flowering trees and shrubs, while the ammonites and belemnites disappeared from the seas.

85 Britain from space 165 million years ago: forested swamps and warm seas

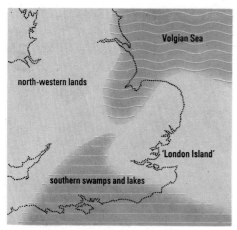

86 Shrunken seas 140 million years ago

Palaeogene Period

The Atlantic was beginning to open right next to the British Isles 60 million years ago. Uplift of the crust around massive volcanoes (figs 93c, 89, 90) caused much of the Chalk in the north-west to be eroded away, while elsewhere muds, sands and chalk continued to settle in shrunken seas. Little by little the crust of England and Wales tilted to the southeast as the North Sea area continued to sag. Now river-sands and marine muds alternately spread over the site of the buried London Island; to the southwest, river-sands and mud-flats extended across the English Channel while the North Sea area was occupied by a muddy sea (front cover). The Antrim and Hebridean volcanoes were waning 55 million years ago, when a deep marine gulf covered the London area and the North Sea, depositing the thick *London Clay* during the Eocene Epoch (fig 92). The London Clay contains a great variety of tropical plant and animal remains. Across palmy shores and crocodile-infested swamps, large mammals and huge land-birds lived untroubled by dinosaurs. By 35 million years ago, oak and beech were growing alongside redwood, cinnamon, palms and magnolia across wide expanses of grass-land, meres and estuaries. Britain was again drifting towards more northerly latitudes.

87 Gault Clay on Greensand, Westerham, Kent

89 Volcanic scenery: Antrim 60 million years ago

90 Lava flows near Giant's Causeway, Co. Antrim

88 Coccoliths magnified a thousand times

91 Chalk cliffs near Swanage, Dorset

92 London Clay pit, South Ockendon, Essex

Continental Drift and the Alpine mountain-building

250 million years ago, Britain was in the middle of a vast supercontinent destined to fragment into Africa, the Americas, Eurasia, Australia and Antarctica. Shallow seas and lakes traced out areas of rifting and subsidence in the continental crust, some of which were later to widen into oceans. When the South Atlantic Ocean opened, Africa drifted east and rotated towards Eurasia, narrowing the ocean gulf of the Tethys Sea (fig 93). Later, the North Atlantic began to open, allowing Eurasia to catch up with the movement of Africa, drastically altering their relative motion. As Africa rotated against Eurasia,

several small continental fragments were crushed and sheared along with the Tethys ocean crust until, inexorably, great mountain chains, including the Alps, arose. The marginal effects of these extensions and compressions on Britain were considerable. As the Atlantic gradually opened up, the rifts around Britain intermittently deepened and filled with sediment. The tensions and compressions of ocean-spreading and mountain-building created an ever-changing continental geography of seas, rivers, lands and lakes. The geological record of these changes in Britain is comprised in rock

textures and compositions, folds, faults and buried erosion surfaces. The most complete records are afforded by persistently subsiding areas such as the North Sea basin. The final separation of the British continental crust from Greenland was preceded by intense volcanic activity located where warps in the crust inter-sected north–south belts of melting below it. Basalt lavas flooded out, followed by rising magma masses forming volcanic centres (figs 93c, 89). Only the last map below (fig 93e) shows the shallow seas and coastlines in continental areas

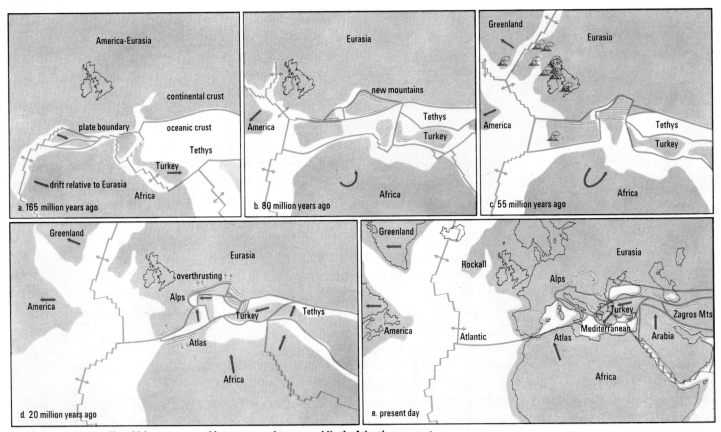

93 A supercontinent splits: Africa rotates, pushing up mountain ranges while the Atlantic opens out

The Alpine folding in Britain came after 200 million years of intermittent crustal warping and rifting associated with the opening of the North Atlantic. Much of the modern structural pattern of Britain was imposed soon after the uplift of the Variscan mountains, though the forces responsible were constrained to act along earlier lines of crustal weakness. These lines bordered persistent upland regions which were, and still are, remnants of the old Caledonian chain. However, the present-day shape of Britain does not everywhere reflect this basic structural pattern: the London Basin, for example, is a temporarily unsubmerged arm of the North Sea basin, while the Cheshire Basin is a structural extension of the Irish Sea basin (figs 10, 94). Some of the strongest rifting movements around Britain occurred at a time of widespread crustal tension around 140 million years ago at the end of the Jurassic Period. Worldwide rates of ocean spreading increased 110 million years ago, inflating the volume of the mid-ocean ridges and thus causing the sea to flood large areas of land. Meanwhile, as the new North Atlantic ocean widened, the tensional rifting around Britain came to an end; a subsiding blanket of sediment now covers the old rifts under the sea (fig 98). The development of the North Atlantic continued after Greenland parted from Rockall Bank, which began to subside after some ten million years of volcanic activity, while uplift rejuvenated old highland areas such as Scotland and the Pennines (fig 96). The final separation of Europe was now complete. Around 25 million years ago, Alpine folding (fig 97) and deep faulting in southern areas of Britain were superimposed on the old crustal pattern in consequence of the varicus continental movements to the south (fig 93d). The broad dome of the Weald and its cross-Channel continuation in the Boulonnais (fig 95), was buckled along the northern flank of the Wessex–Paris Basin. The stage was now set for the final shaping of Britain, while the Atlantic widened by twenty kilometres every million years.

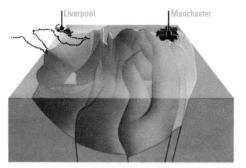

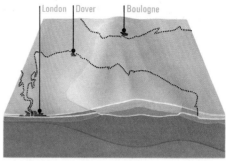

94 An infilled rift: the Cheshire Basin

95 Alpine folding: the Weald Dome

IGS L411

IGS A12222

96 Eroded fault scarp, Giggleswick Scar, N Yorks

97 Alpine folding at Stair Hole, Dorset

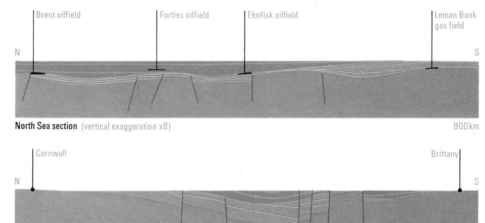

Brent oilfield   Forties oilfield   Ekofisk oilfield   Leman Bank gas field

N   S

North Sea section (vertical exaggeration x8)   900km

Cornwall   Brittany

N   S

English Channel section (vertical exaggeration x10)   200km

Rocks less than 100 million years old   Rocks 100 to 280 million years old   Rocks older than 280 million years

98 Sections across the North Sea (above) and the Channel

Britain emerges

Neogene and Quaternary time

The last 26 million years is a tiny fraction of the Earth's history but it was during this time that the British Isles acquired the shape we know today. Along the widening Atlantic, the western edge of the European plate was stretched and warped. The British mainland and especially the upland blocks of Scotland, Wales, the Pennines, Lake District and southwest England have slowly but repeatedly risen while the floors of the North Sea, Irish Sea and parts of the Channel and the continental shelf to the west have sagged. Uplift in the west and sinking around the North Sea (fig 113) have given much of the country a gentle tilt to the east. Erosion of the rising land supplied sediment which rivers carried into the surrounding seas, but rocks of this time are to be found on land only in East Anglia and a few places in southeast England, which were drowned by a temporary rise in sea level about two million years ago. Here the 'Crags' of Norfolk and Suffolk consist of only a few metres of broken shells and sand swept together by strong currents near the shore, whereas boreholes in the North Sea show that 1200 metres of Neogene sediment piled up on its sagging floor. Certain fossil bivalves and water snails from East Anglia are similar to those inhabiting the mild waters of the Mediterranean today, while others now live in the Arctic. Changes in the fossil population show that the fluctuations from temperate to cold climate which were to dominate the last two million years had already begun. In the west, long-continued erosion by rivers and the sea had worn the folded stumps of the ancient Caledonian and Variscan ranges into extensive plains. Here, crustal upwarping set in and repeated pulses of uplift hoisted these eroded lowlands for hundreds of metres, transforming them into plateaus or 'upland plains' which now dominate the skyline of most of our highland regions. Rivers and glaciers have cut deep valleys into these flat-topped uplands (fig 115), and in places where erosion has been particularly intense, all that is left is a series of isolated peaks which reach approximately the same height. Just when the plateaus were eroded is not at all obvious. Some may have been covered by Chalk or Jurassic formations now worn away, but most were probably planed off during the Tertiary Period. Our highest mountains occur in areas of particularly resistant rocks such as the granites of the Cairngorms and the volcanic rocks of Snowdonia and are surrounded by 'staircases' of lower, younger erosion platforms (fig 114b). These snowy highlands became the breeding grounds of the Ice Age glaciers.

99 Britain from space during the Pleistocene Ice Age at the time of the penultimate glaciation

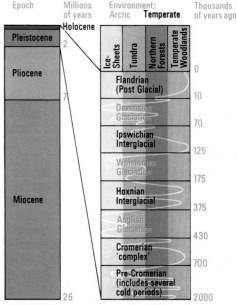

100 Subdivisions of Neogene and Quaternary time

The Ice Age

101 Ice-sheet and adjacent frozen ground

102 Thames Valley in the Ipswichian Interglacial

For reasons still unclear, world climate began to cool in the early Neogene and ice-caps formed at the poles. In the past 2½ million years cold polar waters have advanced southwards more than twenty times and ice-sheets have built up on adjacent land-masses. During these advances, the Gulf Stream flowed eastwards towards Spain and Britain's shores were washed by cold polar currents from Greenland (fig 103). In Britain there were several periods of arctic cold when the average annual temperature fell to −6°C to −9°C (fig 100). All the mountain regions became centres of glacier growth. As arctic conditions intensified, the mountain glaciers coalesced into something like the present-day Greenland ice-cap and ice-sheets spread out across lowland Britain. The earlier glaciations were the most extensive; ice-sheets well over 1000 metres thick reached as far south as North London (fig 104). A vast ice-sheet from the Scandinavian mountains flowed across the swampy North Sea depression and entered eastern England, while another sheet advanced up the English Channel from the west. Land not covered by ice or meltwater deposits showed all the characteristics of permanently frozen ground, the most conspicuous of which was *solifluxion,* the downhill flow of melted surface material as a contorted sludge of soil and rock fragments. In the warm interglacial periods between advances of the ice, the average annual temperature reached or even exceeded that of today. As the climate warmed, the vegetation changed from open grassland to light woodland to luxuriant forest at the climatic optimum, when for example in the Ipswichian Interglacial (fig 102), hippopotamus lived as far north as Leeds. As cold returned, the mixed deciduous forest changed to northern pine and birch, park tundra and finally arctic ice-desert. 10 000 years ago, the last ice-sheet melted from the Scottish Highlands and by 5000 years ago the climate was already warmer than it is today.

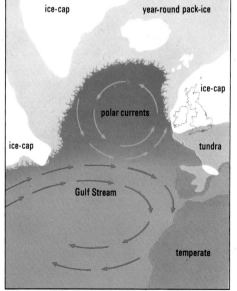

103 North Atlantic in May 18 000 years ago

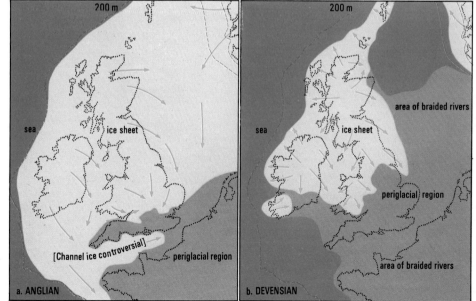

104 Extent and direction of flow of ice sheets in the Anglian and Devensian glaciations

Ice erosion The mountain landscape of Britain, particularly in Snowdonia, the Lake District and the Scottish Highlands, shows clearly the extensive exposure of bedrock and characteristic landforms produced by ice erosion. Around the highest peaks are many steep-walled *corries* – hollow amphitheatres – each marking the birthplace of a glacier. When the glaciers outgrew their corries they flowed down pre-existing river valleys, picking up scree and alluvium and quarrying out bedrock from the floors and sides. All this debris was gripped by the ice and used to scrape, scratch, groove, mould and polish the bedrock (fig 105). These glaciated valleys are now straighter than the original river valleys and more steep-sided so that their cross-sections are U-shaped (fig 106) rather than the V-shape typical of mountain river-valleys. The ice-sheets submerged and scoured all but the highest peaks, creating a smoothed and more rounded landscape.

Ice deposition The lowland landscape of Britain was scraped and smoothed by the ice sheets and then very largely covered by spreads of glacial 'drift' deposited below and at the edges of the advancing ice or left behind when the ice melted. Though the most distinctive drift is *boulder clay* deposited directly by the ice (fig 107), much is water-laid sand and gravel carried by meltwater streams (fig 108). Perhaps the most intriguing glacial deposits are the scattered *erratic blocks* of 'foreign' rock which commonly sit directly on bedrock (fig 109). Occasionally, erratics can be matched with a unique rock outcrop and thus indicate the path of the ice sheet which carried them. Erratics found on the Yorkshire coast, for example, can be matched with rocks in the Oslo area. The same ice that carried these erratics diverted many northerly-flowing rivers including the Thames which before the glaciations followed a north-easterly course through the Vale of St Albans.

105 Ice-scratched rock, Alltsigh, Loch Ness, Inverness-shire

107 Boulder clay, County Fermanagh

108 Outwash gravel, County Down

106 U-shaped glaciated valley, High Cup Gill, Cumbria

109 Perched glacial erratic, Norber, near Austwick, N Yorkshire

Early man and his environment

Fossil remains of early man are extremely rare in Britain, but his flint artefacts are abundant. The earliest authenticated finds are bifacially worked implements and a struck flake from cave deposits at Westbury-sub-Mendip. They were associated with remains of bear, lion, jaguar and sabre-toothed cat dating from the late Cromerian 400 000–500 000 years ago. The oldest human fossil found in Britain, the famous Swanscombe skull-bones, came from ancient Thames gravels dating from the late Hoxnian Interglacial possibly 250 000 years ago. The same deposits have yielded many thousands of handaxes of *Middle Acheulian* type, together with evidence of the use of fire. 'Acheulian man', who lived on into the Ipswichian Interglacial, was seemingly an early form of *Homo sapiens* with a preference for hunting big game in cooler, open country.

By contrast, the makers of the apparently more primitive *Clactonian* implements, found below the level of the Swanscombe skull, were riverside forest dwellers. Though possibly representing merely a different cultural tradition, the Clactonian implements may be the work of a different and older human species, *Homo erectus.* In the warmest period of the Ipswichian Interglacial, man was inexplicably absent. The primitive-looking but intellectually advanced 'Neanderthal Man', a sub-species of *Homo sapiens,* lived in a less severe phase of the last glaciation up to about 35 000 years ago, when he was replaced by modern man, *Homo sapiens sapiens* (Cro-Magnon Man). Whether they actually coexisted in Britain is uncertain. Both inhabited caves and hunted reindeer, mammoth and woolly rhinoceros across open grassland and tundra.

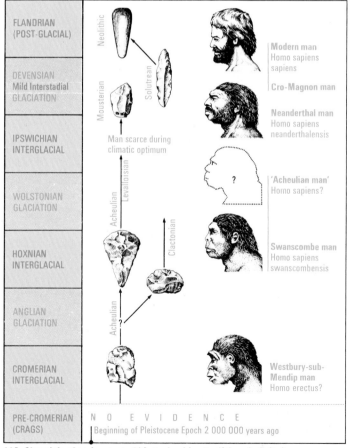

110 Chronicle of early man and his toolmaking traditions in Britain

111 Animals hunted for food and skins by early man

112 Lakeside campsite of 'Acheulian man' in the Hoxnian Interglacial

The shaping of the scenery

As we have already seen, the broad features of Britain's highland and lowland were blocked out by uplift, tilting and erosion during the Tertiary Period before the Ice Age (fig 113), but all the detail of our present scenery has been shaped within the last half million years. The mountains of Scotland, Ireland, northern England, Wales and the southwest still show the 'grain' of the Caledonian or Variscan foldings. Although upland erosion plains have cut across these ancient structures (fig 114a), outcrops of resistant rock often produce belts of rugged country while softer beds or the shattered rocks of fault-zones guide the valleys. The Irish highlands are more fragmented by erosion.

The softer Mesozoic sedimentary layers of the Midlands, eastern and southern England are only gently tilted or folded. Here tougher limestones (including Chalk) and sandstones stand up as escarpments which are slowly worn back as the softer clays are eroded into broad vales. Isolated

113 Tilting of Britain in the past 50 million years

a   Surface of erosion cutting rocks of various ages and structures

b

1st uplift accordant summits   2nd uplift   3rd uplift   4th uplift

Erosion surfaces resulting from repeated uplift of the land

114 Erosion surfaces seen as (a) accordant summits and (b) stepped scenery

F. W. DUNNING IGS

IGS A10149

115 Accordant summits from Bwlch y Groes, Gwynedd

116 Submerged forest, Neolithic Forest Bed, Bexhill, Sussex

hills and uplands like the Mendips, Malverns, Wrekin, Charnwood Forest and Cannock Chase are high spots of old hard Palaeozoic and Precambrian rocks jutting through holes worn in the carpet of younger sediments.

Many features of the scenery of our coasts and river valleys result from changes in the levels of land and sea that occurred during the Ice Age and are still going on. The causes are complicated and include local uplift, sinking and warping of the land and also world-wide changes in the level of the sea surface. Patches of rounded beach pebbles on the Chilterns and the North Downs containing sea shells about a million years old show that the shore was then 200 metres above present sea level. During subsequent glacial periods, sea level fell to at least 135 metres below its present height as moisture evaporated from the ocean surfaces throughout the world and then fell as snow to build the gigantic American and European ice-sheets. Large areas of the continental shelf became dry land and Britain was joined to the mainland of Europe. The rivers of southern England, then tributaries of the Rhine and the Seine, cut deep valleys to the sea. As the climate improved and the ice melted, the sea advanced again, drowning the valleys (fig 117) and form-ing estuaries like those of the Thames, Severn and Southampton Water as well as the steep-sided *rias* of Milford Haven and Falmouth Sound. The flooding of former glaciated valleys produced the fjord coasts of western Scotland and Ireland. The tree stumps of the submerged forests found in places around our coasts (fig 116) were covered by the rising water only 6000 years ago. As the British ice-sheet melted, its enormous load, which had been weighing down this part of the Earth's crust, was removed. At the rate of a few millimetres a year, the land recovered and in places is still rising. Rocky, wave-eroded shores, some covered by gravel, sand and shells, were uplifted far above the highest tides to become raised beaches (fig 118). As rivers adjusted their flow to changing sea levels, terraces formed on the valley sides.

117 Drowned valley or *ria*, Salcombe, Devon

118 Raised beaches, Dunure, Ayrshire

Recent changes

When the present temperate phase began 10 000 years ago, the ice, melting and retreating from Britain, provided torrents of water which sought the sea across the exposed continental shelf. The sea rose rapidly, creating a widening gap in the land bridge which connected Britain to the Continent, and submerging low-lying coastal regions, to reach its present level some 5000 years ago. Since then, Britain's coastline has receded even further. Not only are some stretches of coast sinking (fig 119), but all the coast is under continuous attack by the sea. Coasts of soft rock, such as boulder clay, are being eroded at sometimes disastrous rates (fig 120) by direct attack coupled with extensive landslipping. Some land is being gained in sheltered bays where estuarine deposits, colonised by salt marsh plants, develop into coastal flats. Such areas are often artificially drained behind sea walls, as for example Romney Marsh.

On land, the rivers in their lower reaches responded to the rise in sea-level by filling their over-deepened channels with alluvium. Though the rivers have cut into the glaciated landscape, its characteristic features, the U-shaped valleys, the undulating lowland plains of boulder clay, appear almost untouched. Nevertheless the land surface is being steadily worn down, mainly by the action of rivers (fig 122), and also by the imperceptible mass movement of loose rock and soil down every hillside. Man's activities have profoundly changed the appearance of the land-scape. The clearance of natural forest begun in Neolithic times also increased the rate of erosion. Peat-cutting in historic times led to the flooding of the Norfolk Broads. Since the 19th century the exploitation of Britain's surface rock resources has grown to affect large areas of the countryside. Huge quantities of ironstone, limestone (fig 121), road aggregate, brickclay and china-clay are extracted from quarries and pits. Over 100 million tons of sand and gravel are dug each year for concrete, much of it for motorways which carve through hills and fill valleys.

RISING

+3
+2
+1

−1

−1

0

This area has sunk by 6m in the last 6500 years

−2

−2

SINKING

—— eroding coastline
—— land gaining
—— stable coastline
⌄ contours at 1mm per year intervals

119 Present-day coastal changes and crustal movements

IGS A6943

120 Collapsed houses on an eroding cliff, near Lowestoft, 1936

IGS L930
(OPPOSITE: KENNETH SCOWEN)

121 Hillside quarried for limestone, Wirksworth, Derbyshire
122 (opposite) Symonds Yat: a valley eroded in Old Red Sandstone by the R. Wye